Education 140

我喜欢跳啊跳

I Like to Move It, Move It

Gunter Pauli

[比]冈特·鲍利　著

[哥伦]凯瑟琳娜·巴赫　绘

何家振　译

上海遠東出版社

特别感谢以下热心人士对童书工作的支持：

匡志强　宋小华　解　东　厉　云　李　婧　庞英元
李　阳　梁婧婧　刘　丹　冯家宝　熊彩虹　罗淑怡
旷　婉　王靖雯　廖清州　王怡然　王　征　邵　杰
陈强林　陈　果　罗　佳　闫　艳　谢　露　张修博
陈梦竹　刘　灿　李　丹　郭　雯　戴　虹

目录

Contents

蜜蜂们一边在蜂巢里休息，一边观看村子里举行的舞会。乐队演奏萨尔萨、梅伦格、桑巴，甚至恰恰舞曲。不久，人们就开始随着舞曲跳起来了。

一只工蜂观察了一会儿，说道："人类很喜欢跳舞哇。"

The bees are resting in their hive and watching a party unfold in the local village. The band plays salsa, merengue, samba and even the cha cha cha. It does not take long for people to start dancing to the rhythm of the music. "People do like to move it," observes one of the worker bees.

蜜蜂们在蜂巢里休息……

The bees are resting in their hive ...

只要曲子一响……

As soon as I hear a tune ...

“是的。”蜂王被人们快速舞动的身体逗乐了，“你们看，音乐把语言、感情和动作汇集在一起了！”

“我太喜欢了。只要曲子一响，我的内心就能感受到，翅膀也随之扇动起来。”

“确实如此，只要音乐响起，所有人都有同样的感觉。音乐让我们团结在一起。它巩固了我们的关系，支配了我们的情绪。”

“Indeed,” replies the queen bee, amused by the sight of fast moving bodies. “You see, music brings together languages, feelings and movements.”

“I just love it. As soon as I hear a tune, I feel it in my heart and immediately my wings are ready to flap.”

“Absolutely, when there is music playing, everyone is kind of feeling the same way. It makes us bond. It cements ties and determines our mood.”

“只有人类才跳舞吗？”工蜂好奇地问。

“你忘了我们蜜蜂会跳摇摆舞吗？”

“当然不会，我怎么会忘记我们的摇摆舞呢？我们用跳舞来告诉我们的同伴去哪里寻找食物。如果有一只蜜蜂掉队，比如说被风吹走了，我们就摇摆起来，然后所有蜜蜂就又被召集到一起。”

“Are people the only ones who dance?” the worker bee wonders.

“Did you forget that we bees perform a waggle dance?”

“Of course, how could I ever forget our waggle dance? We dance to tell our buddies where to find food. If one of us gets lost due to strong wind for instance, we waggle and bring everyone together again.”

……我们蜜蜂会跳摇摆舞吗？

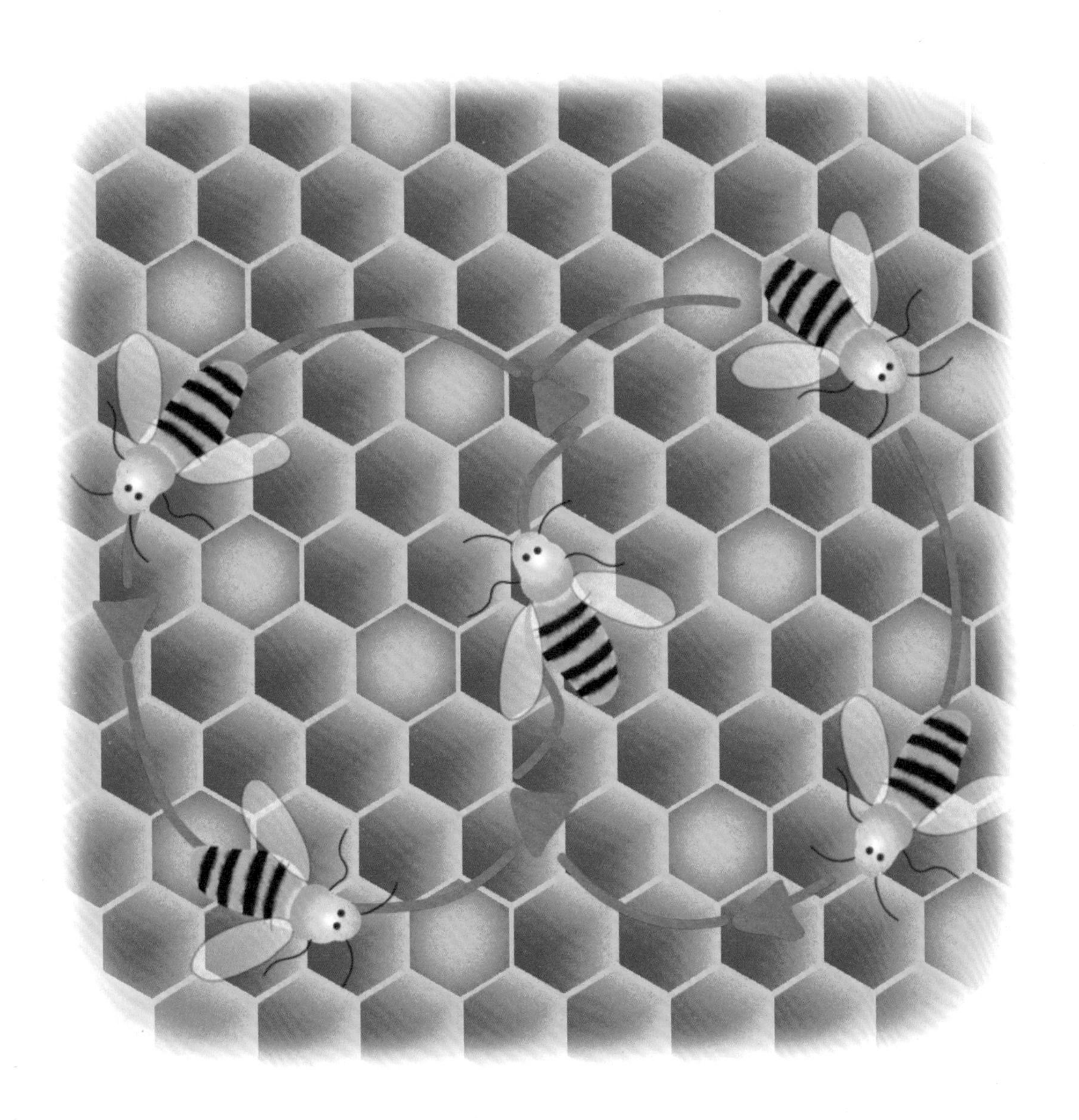

... we bees perform a waggle dance?

我们的朋友凤头鹦鹉也具有音乐节奏感。

Our cockatoo friends have great rhythm too.

“我们的朋友凤头鹦鹉也具有天然的音乐节奏感。”蜂王补充道，“那些鸟儿可以响应人类创作的音乐。我见过它们跟着皇后乐队演唱的《又打倒一个》的节奏狂舞。”

“Our cockatoo friends have a wonderful natural sense of rhythm too,” the queen bee adds. “Those birds respond to music made by people. I have seen them go wild to the beat of ‘Another One Bites The Dust’, a song by a band called Queen.”

“是的。我知道那种鸟儿移动和晃动的能力令人难以置信。但是我觉得，最疯狂的舞者是圆尾娇鹟。这个名字的意思是‘跳跃者’。这种鸟生长在哥伦比亚和厄瓜多尔丛林，表演的迈克尔·杰克逊太空步非常棒！即使迈克尔本人也会为其感到骄傲。”

“Yes, I know that birds are incredible movers and shakers. But for me the craziest dancer is the manakin. The name means ‘jumper’. This bird from the jungles of Colombia and Ecuador performs Michael Jackson’s moonwalk dance! Even Michael would be proud.”

最疯狂的舞者是圆尾娇鹟。

... the craziest dancer is the manakin.

……孔雀蜘蛛的舞蹈……

… the dance of the peacock spider …

“你见过孔雀蜘蛛的舞蹈吗？那是一种腹部有彩绘的蜘蛛。你知道吗，雌蜘蛛会与雄蜘蛛一起跳舞，以鉴别他作为结婚对象是否足够健康。”蜂王问道。

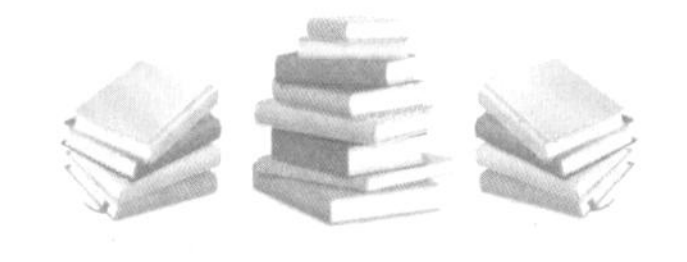

“Have you ever seen the dance of the peacock spider, that spider with the painted belly? You know the lady spider dances with the male to see if her partner is healthy enough to marry?” asks the queen bee.

“虽然我喜欢画得五颜六色地出席舞会，但是我认为，仅仅凭跳舞来判断你未来的伴侣并不公平。很多人能成为生活中的好伴侣——尽管他们根本不会跳舞。他们的内心仍可以有美好的韵律。”

“你说得太好了。谢谢你那样说。你知道谁是华尔兹冠军吗？”

“While I love the idea of getting all painted to go to a party, I do not think it is fair to judge your future partner only by the way he dances, though. Many could be great partners for life – even if they cannot dance at all. They can still have great rhythm in their hearts.”

“You are so right. Thank you for saying that. Do you know who the champion of the waltz is?”

凭跳舞来判断你未来的伴侣并不公平。

Not fair to judge by the way he dances.

那一定是奥地利人……

That must be an Austrian ...

“华尔兹？那一定是奥地利人，那里是华尔兹乐曲诞生并发扬光大的地方，这都要归功于作曲家弗朗茨·约瑟夫·施特劳斯。”

“华尔兹冠军是淡水水藻！很令人意外吧？”

“恕我直言，亲爱的蜂王——这真是难以置信。”

“The waltz? That must be an Austrian, as that is where that kind of music was first composed and became so popular, thanks to the composer Franz Josef Strauss.”

“Surprise: the champion of the waltz is fresh water algae!”

“With all due respect, my dear Queen – that is difficult to believe.”

“好，下次你飞过水面时，仔细观察水藻如何相互拥抱，彼此围绕着旋转。那样你就会意识到它们真的是华尔兹冠军。”

“真是让人惊奇，”工蜂说道，“是不是其他的淡水水藻那时候都在唱‘我喜欢跳啊跳’呢？”

……这仅仅是开始！……

“Well, next time you fly over the water, take a closer look and see how algae embrace each other and spin around, orbiting one another. Then you will realise that they are indeed the waltzing champs.”

“That makes me wonder,” the worker bee says, “do all the other fresh water algae then sing, ‘I like to move it, move it’?”

... AND IT HAS ONLY JUST BEGUN! ...

……这仅仅是开始！……

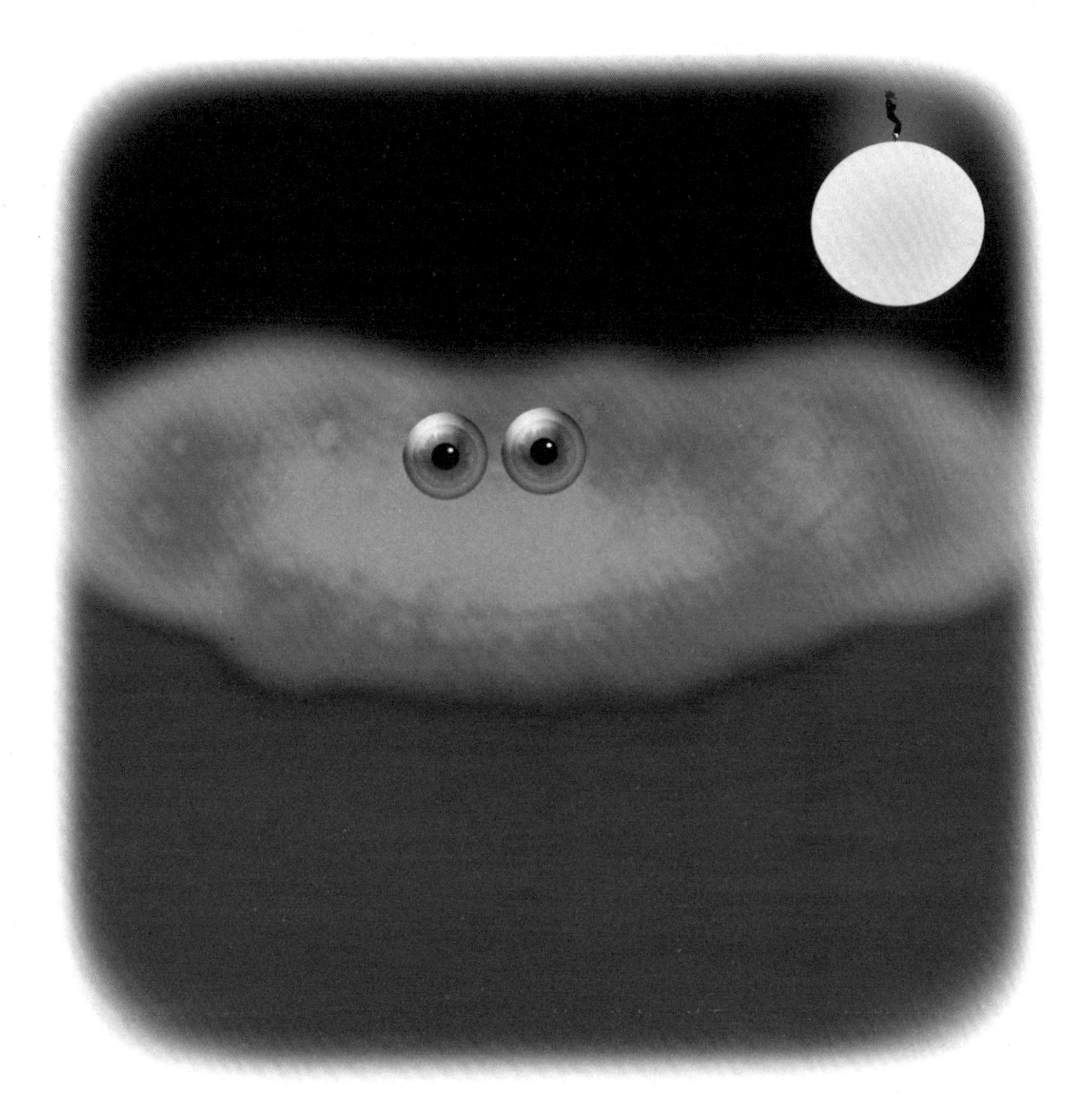

... AND IT HAS ONLY JUST BEGUN! ...

Did You Know?

你知道吗？

Music influences our emotions. Happy music seems to make us even happier, sad music seems to make us feel melancholy. Music has the power to make you move and dance, by yourself or with friends.

音乐可以影响我们的情绪。快乐的音乐会使我们更快乐，悲伤的音乐让我们感到更悲伤。音乐具有一种让人动起来和跳起来的力量。

Music consists of modulations, pitch, intensity, tempo and rhythm. The acoustics of music can be described mathematically. Measurement of time, frequency and movement in dance finds analogies in geometry.

音乐由转调、音高、强度、节拍和音律组成。音质可以用数学方式表达。跳舞的时间、频率和动作的测量都能在几何学中找到相似之处。

According to the experts there are more than 300 genres of music practiced in the world today. Latin American dance styles are dominating the world including the bachata and merengue (Dominican Republic), cha cha cha, rumba and mambo (Cuba), salsa (Caribbean), and samba (Brazil).

根据专家的说法，世界上有超过 300 个音乐流派。拉丁美洲的舞蹈风格主导着世界，包括芭恰塔、梅伦格（多米尼加），恰恰、伦巴、曼波（古巴），萨尔萨（加勒比）和桑巴（巴西）。

The Chinese, Indians, Egyptians and Mesopotamians have studied the mathematical principles of sound. The expression of musical scales, in terms of numerical ratios and particularly the ratios of small integers, has been studied in ancient Greece.

中国人、印度人、埃及人和美索不达米亚人研究过声学的数学原理。早在古希腊就已经有人研究用数值比率，特别是小的整数比率来表达音节了。

诺贝尔奖获得者卡尔·冯·弗里施提出了“蜜蜂的摇摆舞是告知彼此哪里有食物”的理论。科学家们用雷达证明了这一理论的真实性。

Nobel Laureate Karl von Frisch theorised that the waggle dance of bees is a way to tell each other where to find food. Scientists using radar confirmed that this is true.

孔雀蜘蛛原产于澳大利亚。雌性通过跳舞来判断雄性是否足够健康，可以成为理想的生育伙伴。没能通过这种检验的雄性蜘蛛将会被杀死并被吃掉。

Peacock spiders are native to Australia. The purpose of their dance is for the female to determine if the male is healthy enough to become the ideal reproductive partner. Those males who do not pass this test are killed and eaten.

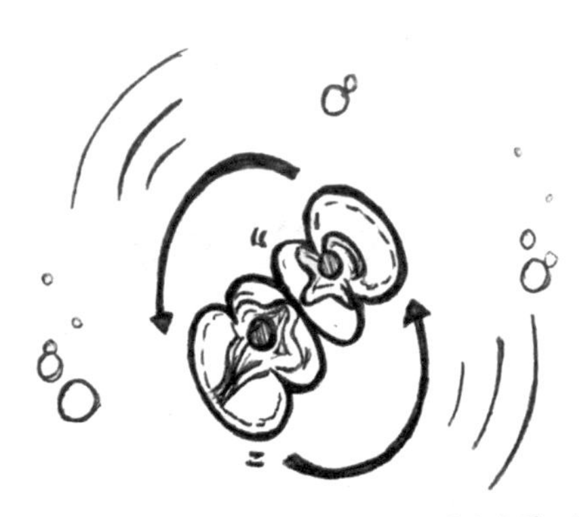

Researchers at Cambridge University discovered that fresh water algae use movements like that of the waltz and the minuet to reproduce. Members of a colony either orbit around each other as in a waltz, or move back and forth as in a minuet. Both instances lead to clumping and reproduction.

剑桥大学的研究人员发现，淡水水藻通过像华尔兹和小步舞那样的运动繁殖。住在一个菌落的水藻像跳华尔兹那样围绕着对方旋转，或者像在小步舞中那样前后移动。这两种方式致使水藻丛聚和繁殖。

The colourful manikin bird native to Colombia and Ecuador has a mating dance that includes moves almost identical to Michael Jackson's moonwalk.

原产于哥伦比亚和厄瓜多尔的侏儒鸟的求偶舞，包括了与迈克尔·杰克逊的太空步几乎完全相同的动作。

Think About It 想一想

Would you expect birds to dance to the rhythm of music, or do you consider this something only humans could do?

你愿意相信鸟儿们可以踏着音乐的节拍跳舞，还是认为只有人类才会这样做？

Are you dancing to feel good, to exercise or to be noticed?

你跳舞是为了寻求美好的感觉，还是为了锻炼，或者是为了引人注意呢？

If bees are showing each other the way to food by waggling, can waggling then be rightfully considered a dance?

如果蜜蜂通过摇摆指引彼此的觅食之路，那么还应该把蜜蜂的摇摆看作跳舞吗？

How can algae waltz or minuet, for these single cells do not have legs to move around? How do these tiny species create movement?

水藻是怎么跳华尔兹或小步舞的呢？这些单细胞生物明明没有腿，却能四处飘动。这些微小的物种是如何运动的呢？

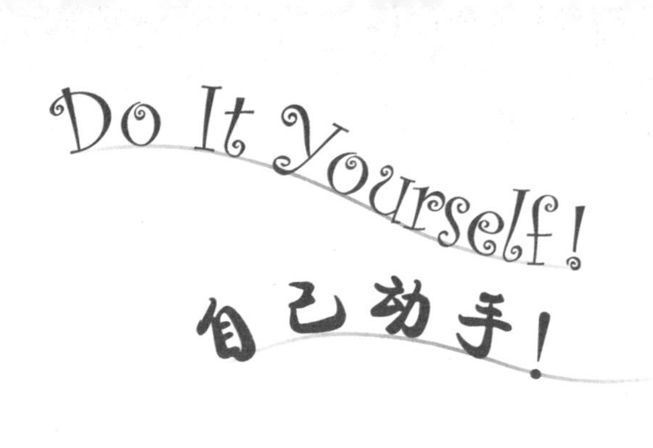

Have a look at Michael Jackson's moon dance and watch the video, of the manikin bird dancing. Now, if this bird can do the backward steps, so can you! First try this without the music, just to get the correct body movements, and then try it again with the music in the background. Who in the class is the best moon dancer?

看一下迈克尔·杰克逊的太空步，再看一下侏儒鸟跳舞的视频。如果这种鸟能够向后滑步，那么你也可以做到！先不放音乐，只做正确的身体运动。然后再试着跟着背景音乐做。谁是你们班里最好的太空步舞者?

学科知识

Academic Knowledge

生物学	给植物播放古典乐，可以促进它们生长；DNA或者蛋白质序列可以变为音乐；如何将来自兰花草、秋海棠、橄榄树等植物的生物电流转化为音乐作品；给葡萄藤播放莫扎特音乐使会葡萄产量提高50%；跳舞可以使身体变得灵活敏捷，并使平衡性和体态变好。
化　学	音乐促进大脑产生更多的神经递质多巴胺和血清素（使大脑产生“幸福感”的化学物质）。
物　理	声音是由物体振动产生的，这种振动让作为介质的水或者空气也发生振动；振动形成了我们可以听到的纵向声波，声波的波长和速度决定了声音的频率（也称为音调）；在乐器（比如吉他）中，弦的长度和粗细决定不同的音高；一个八度音阶由12个半音组成。
工程学	建筑声学应用于营造室内最好的音效，控制办公室内、铁路等的噪音。
经济学	音乐和舞蹈是繁荣的经济成分；为音乐和舞蹈消费的倾向在增强；从1969年起，伍德斯托克音乐节门票价格涨了60倍，乐队的收入也增加了20倍，然而人们变得越来越不愿为音乐本身买单；音乐行业的盗版行为致使艺术家收入减少。
伦理学	美德与艺术鉴别力的区别：使人成为艺术鉴赏家的技术能力，并不能保证其具有美德。
历　史	人类在学会说话之前就学会了歌唱和舞蹈；约翰·塞巴斯蒂安·巴赫的复调音乐使音乐发生了革命性变化。
地　理	音乐和舞蹈是世界各地区的传统和文化的一部分，这是真正的全球性活动。
数　学	毕达哥拉斯观察了琴弦振动与和声之间的基本数学关系，从几何图形中得到了声调符号；波长与频率的反比关系；现代作曲家使用2的12次方根谱写乐谱，以保证音调之间的均衡；音乐几何学，相似性是几何学的定义性特征，而旋律则是声音的相似性排列。
生活方式	在小学阶段参与音乐和舞蹈的表演，可以降低日后患糖尿病和心脏病的风险。
社会学	跳舞和听音乐可以将人们聚集在一起，使人们以非语言的方式沟通交流，相互影响。
心理学	音乐、舞蹈与情绪有重要联系，同时，舞蹈者的情绪也影响着他们的动作；音乐标志着人的个性，但音乐偏好的变化并不改变人的个性；运动疗法——接近、处理和调节情感。
系统论	音乐和舞蹈（声音和动作）在很多学科中都结合在一起，包括计算机科学、播音工程、物理学、声学工程、电机工程和电子工程、体育、舞蹈艺术。

情感智慧

Emotional Intelligence

蜂　王

蜂王看到那些跳舞的人后被逗笑了；她的评论很有洞察力，给工蜂们一种归属感。她在回答工蜂的问题时提出了一些问题，并举出例证，鼓励工蜂们更仔细地观察鸟类的行为，了解事情的多样性。她用一种独有的同理心去交流，把工蜂们当成家庭成员，很尊重她们。蜂王很诙谐，列举了一些世界各地动物们的惊奇之处，并且再次证实了“人不可貌相”，不能仅仅凭他们跳舞的方式来判断，而是要看他们怎样分享内心的感受。当遭到怀疑时，她依然保持冷静，指出了获取知识的途径，从而慢慢萌生更多地了解他人的愿望。

工　蜂

工蜂是一个劳动者，很尊重蜂王。她愿意与蜂王分享自己的感受，她在向蜂王提问时感到很自在。当蜂王以问题作为回答时，工蜂热情洋溢地展示了她的专门技能。她自我克制，不过多地显露自己的知识，但是她乐于分享信息和回答很多问题。她还分享了一个关于通过跳舞判断他人的观点。她们的对话发展到令人惊奇的地步，工蜂不怕冒犯地告诉蜂王自己很难相信她所说的。在蜂王的指导下，工蜂展示了一种幽默感。

艺术

The Arts

舞蹈和音乐紧密相连。选择3个传统音乐流派：亚洲一个，拉美一个，非洲一个。把音乐与舞蹈匹配。哪种音乐流派对于学习跳舞更容易？哪种音乐流派具有更硬朗的舞蹈动作？哪种音乐流派让你在跳舞时动作有更大的自由度？观察一下，不同文化传统里的乐器有哪些不同。我们如何解释拉丁美洲和非洲的音乐与舞蹈有更多的共同点？

思维拓展

Systems: Making the Connections

音乐和舞蹈是现代社会不可分割的组成部分。人类学家指出人类在具备语言能力之前就具备了音乐和舞蹈的能力。声乐艺术和身体运动是我们日常生活的重要部分。现代的生活习惯让人不得不久坐，孩子们在很小的年纪就花了太多的时间静坐。让人更焦虑的是孩子们的座椅没有随着他们身体发育而调整。医学证实年轻时缺乏锻炼的人更容易罹患肥胖等食物紊乱症，并随时间推移成为糖尿病、心脏病高危人群。将音乐和舞蹈从学校课程里删除是目前的趋势。但是如果作为课外活动，无论男孩还是女孩都可能会喜欢上唱歌跳舞，喜欢探索通过节奏、运动或声音表达自我的独特方式。音乐不只是让人们身体想动，美好的旋律与和谐的音调对我们的精神状态也起到积极作用。欣赏或者演奏音乐不仅让人愉悦，也对我们的身心健康有利。音乐对人类所具有的积极作用同样适用于动物、植物，甚至藻类。音乐是一切生命非常重要的一部分——音乐使葡萄长得更大，兰花随着音乐发出脉冲，藻类利用类似华尔兹的运动繁殖。看起来只有很少的物种是不动的，几乎所有的生物体都可以通过肌肉（在酸度方面的区别）或者借助风（在密度方面的区别）运动。运动无所不在。通过更多的运动，特别是基于世界各地各种不同韵律的运动，孩子们不仅可以发育得更健康，还能学到很多其他地方的文化，而且会更有幸福感。快乐而健康地活着——这似乎是所有父母对其后代的期盼。

动手能力

Capacity to Implement

舞蹈离不开音乐。找到3个你最喜欢的音乐流派，并测量一下每分钟的节拍数，观察舞蹈，算出每个节拍有多少舞步。请注意脚步只是舞蹈的一部分，还有胳膊和身体动作。你能做出脚步、身体动作与节奏之间关系的示意图吗？这样你就从艺术转移到几何学了。

故事灵感来自

This Fable Is Inspired by

卡尔·冯·弗里施

Karl von Frisch

卡尔·冯·弗里施是一位奥地利科学家，他后来成为一位关注动物行为的专家。他因为关于蜜蜂的若干发现而声名鹊起。1956 年，他第一个揭示蜜蜂有彩色视觉、光感知能力、味觉，以及利用太阳自我导航的能力。他发现蜜蜂体内有记录时间的生物钟，能够辨别蜂巢水平，同时也具有感知垂直位置的能力。冯·弗里施观察了蜜蜂的摇摆舞，并认识到这是一种用于分享食物位置信息的语言。早在 1927 年他就在《跳舞的蜜蜂》一书中提出了这种理论。他的第三个重要发现是：蜂王通过排放信息素维持蜂巢里的秩序。信息素是用来吸引在外的雄蜂回到蜂巢来交配的。1973 年他与康纳德·洛伦茨一起获得了诺贝尔生理学或医学奖。

图书在版编目(CIP)数据

冈特生态童书.第四辑:修订版:全36册:汉英对照/(比)冈特·鲍利著;(哥伦)凯瑟琳娜·巴赫绘;何家振等译.—上海:上海远东出版社,2023
书名原文:Gunter's Fables
ISBN 978-7-5476-1931-5

Ⅰ.①冈… Ⅱ.①冈… ②凯… ③何… Ⅲ.①生态环境-环境保护-儿童读物—汉、英 Ⅳ.①X171.1-49

中国国家版本馆CIP数据核字(2023)第120983号

著作权合同登记号图字09-2023-0612号

策　　划 张　蓉
责任编辑 曹　茜
封面设计 魏　来李　廉

冈特生态童书
我喜欢跳啊跳
[比]冈特·鲍利　著
[哥伦]凯瑟琳娜·巴赫　绘
何家振　译